Space Elevator Concept of Operations

International Space Elevator Consortium

Fall 2012

Robert E. "Skip" Penny Jr
Peter A. Swan, Ph.D.
Cathy W. Swan, Ph.D.

**A Study for Progress in
Space Elevator Development**

Space Elevator
Concept of Operations

Published by Lulu.com
pete.swan@isec.org

Cover image by Frank Chase
chasedesignstudios.com

ISBN 978-1-300-81547-1

Printed in the United States of America

PREFACE

The International Space Elevator Consortium is continuing to reach out to promote a better understanding of the space elevator arena. Each year the ISEC focuses on a selected topic over many activities. This includes, during the year-long topic, discussions through its annual conference, papers in its journal [CLIMB], and a study assessing the topic by discipline experts. A focus on operations was selected by the International Space Elevator Consortium [ISEC] for 2012. The "Space Elevator Concept of Operations" was chosen as the appropriate focus at the International Space Elevator Conference (held in Seattle in August) with additional concentration during concurrent release of the electronic version of CLIMB. This combination of activities focused upon one topic resulting in a year-long activity assessing the major components and concept of operations for a space elevator transportation infrastructure. This study report presents the current thinking on how a fully developed commercial space elevator will operate. It draws on the experiences of the study participants and the authors who have over eleven decades of major space system acquisitions and operations experience. This 2012 year-long activities and this study were sponsored by the ISEC whose mission is:

> "...ISEC promotes the development, construction and operation
> of a space elevator as a revolutionary and efficient way to space
> for all humanity ..."

Ted Semon
President, ISEC
www.isec.org

Table of Contents

Executive Summary

T he International Space Elevator Consortium conducted its 2012 study addressing the Concept of Operations for a future Space Elevator Infrastructure. This report presents the findings and conclusion from the authors and participants in the study.

Finding # 1:   While the development of Space Elevator tethers and climbers is a daunting task, their operation will leverage 50 years of satellite operations experience. The climber is essentially a satellite just like the thousands that have been launched to date. The classic "telemetry, tracking, and command" functions (TT&C) for the climber and tether operations will be the same as those for today's satellites. Climbers will be in constant contact with the operations center, capable of autonomous operation, and execution of instructions. The GEO Node will be a satellite capable of remote controlled and autonomous operations as the hundreds that have been functioning in the geosynchronous belt for years. The Apex Anchor will also be a satellite and its operation is even simpler than the GEO Node.

Finding # 2:   The Marine Node, comprised of the Floating Operations Platform and Ocean Going Vehicles, will leverage hundreds of years of deep-ocean off-shore drilling operations. Maintaining platform position, on-loading and off-loading of supplies and personnel, providing living, recreational, and maintenance facilities are exactly the activities off-shore drilling operators do today.

The Headquarters and Primary Operations Center will be the principle location to ensure robust operations across the multiple centers are synchronized, to include: The Climber Ops Center, Tether Ops Center, Floating Ops Center, GEO Node Ops Center, and the Enterprise Operations Center which includes the Business Ops, Transportation Ops, and the Corporate Headquarters. Each of these Ops Centers supports, or leads, one of the fourteen functions identified within the study to be conducted at typical satellite operations centers.

Finding # 3:   The operations cost, for a pair of space elevators, seems to be reasonable for a business of its projected magnitude.

The authors conclude:

Operations for a Space Elevator
Have No Showstoppers
Have Reasonable Costs &
Meets the Challenge

1 Introduction

The modern day space elevator concept was refined by Dr. Brad Edwards [SE] during the last part of the 20th century. With over 40 years of experiences in space, the industry was ready to support the idea that an inexpensive transportation infrastructure to space could be constructed as soon as the materials were available. The resulting operations would be a mix of historic maritime transportation approaches and unique day-to-day space operations. It seems reasonable that an operations concept could be structured around these two mature industries and costs could be extrapolated to estimate operations and maintenance approaches.

This report addresses initial commercial operations of a space elevator pair with robotic climbers. This report has been developed to help define a starting point for an initial space elevator infrastructure. It is assumed that there are two space elevators in place to ensure continuation of our escape from the gravity well. It also assumes that a sufficient number of climbers are available for delivering spacecraft and other payloads to orbit and. if required, return them to Earth. In addition, this report is designed to be the initial operations concept from which many improvements will occur as future knowledge and experience drives infrastructure concept revisions. The description of a concept of operations, including a quick look at the transportation to space infrastructure, is broken into four sections:

Part I: Mission Description: This portion of the ISEC report deals with the description of individual missions for each major segment of operations. In the space industry, the first step is to refine the needs of the customer and then identify the requirements leading to the development of a program. This study breaks out major segments of the space elevator and discusses the missions which will meet the needs of operators and the requirements that could enable construction. The infrastructure is shown and the philosophies and constraints are addressed.

Part II: System Characteristics: This portion of the document lays out major segments and analyzes their physical characteristics. Each of the segments is discussed as they will be operating in a commercial setting. The concept is that once the infrastructure is laid out, the operations concepts can be structured showing individual activities at each node and operational functions to be performed. Images of the potential structures are shown to illustrate the knowledge level at this time and to help place each major segment in perspective. The major segments of a space transportation system, called a space elevator, are:

- Apex Anchor
- GEO Node
- Space Elevator
- Climber Element
- Marine Node
- Communications Network
- Transportation
- Headquarters

Part III: System of Systems Operations: This section lays out various components of operational activities. There must be a communications network that ties each segment together with high speed data and voice 24/7/365. As the various components are melded together, the majority of operations can be conducted at Headquarters and Primary Operations Center. There will be requirements for remote locations, specifically the Marine Node, where people and facilities will be needed to support day-to-day operations. In the future, there could even be a requirement for people at the GEO node to conduct operations. The last portion of this section breaks out staffing and costing, leading to an estimate of the annual costs for a space elevator infrastructure.

Part IV: "A Day-in-the-Life":   This last section lays out the concept of how one day unfolds for an individual at the Marine Node. This "story-line" approach is based upon the history of space operations as it enables everyone in the development of a future system to understand the needs and complexity of day to day operations. In this case, the transportation of personnel to the south Pacific eats into the monthly schedule of operations. Remote operations are always expensive in terms of support and complexity; however, with the deep-sea oil industry leading the way in operations at sea, Maritime Node operations can be extrapolated fairly easily.

Findings and conclusions from this report are laid out so that decisions can be made for future infrastructures. Hopefully, this analysis will help developers of the space elevator to understand the needs of operators who will have to make components work together efficiently and successfully.

The report utilizes the approaches and formats described in *Cost Effective Space Mission Operations* [CESMO] and *Applied Space Systems Engineering* [ASSE].

Part I – Mission Descriptions
Why a Space Elevator?

The needs of the space elevator are relatively simple:

- Chemical rockets cannot get us to Low Earth Orbit (LEO), Geo-Synchronous Earth Orbit (GEO), and beyond economically: The physics of lifting off with rockets is not energy efficient…and is bad for the environment.
- A less expensive (easier) access to space would mean the "*Space Option*" will become more attractive for enabling solutions to Earth's current limitations: There are many exciting missions waiting for an infrastructure that can lift to GEO and beyond routinely, inexpensively, and safely.

Key changes that will revolutionize the space industry are:

- **Routine:** Space access will become boring and routine with lift-offs occurring every day with 20 ton tether climbers.
- **Price:** The price for a payload to be delivered to GEO will be below $ 500/kg. This change from $ 20,000/kg will alter the clientele for space lift and open up space to businesses that are not even around today.
- **Safety:** Elevators are inherently safer as compared to the dangerous practice of mounting valuable payload on top of huge explosive tanks.
- **Delivery Dynamics:** Space elevators will have vibrations in the region of cycles per day and shock loads of marshmallows dropping into a pool instead of explosive potential and the shake, rattle, and roll of rocket liftoff.

2 Missions

A Space Elevator's mission is to deliver spacecraft to LEO, GEO, and beyond. At the Marine Node (where the space elevator terminus is attached), spacecraft payloads will be mated to a climber that will lift them to drop-off altitudes.

- GEO payloads will be released and will maneuver to the desired GEO slot and desired inclination.
- LEO and MEO payloads will be released below GEO where the spacecraft will circularize its orbit to the desired apogee/perigee and shift to mission inclinations.
- Beyond GEO payloads will be lifted, possibly with a different climber, to the pre-determined drop-off altitude where they will be inserted in a mission trajectory.
- End-of-life satellites and space debris objects may be returned to earth or placed in disposal orbits.

Figure 1 shows the layout of the Space Elevator Infrastructure.

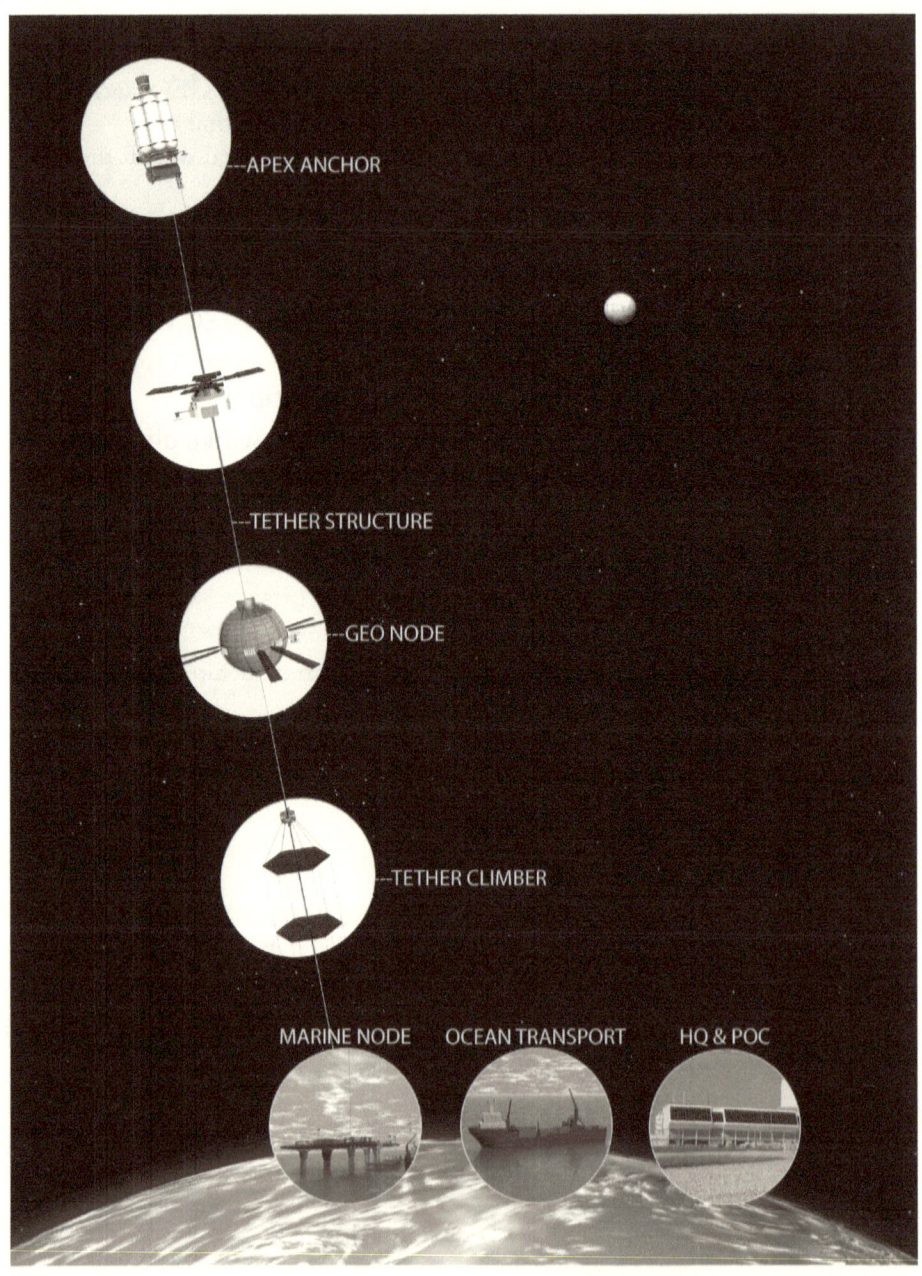

Figure 1 Space Elevator infrastructure [chasedesignstudios.com]

2.1   Climber Mission

The mission of the climber is simple: deliver customer payloads to desired locations safely and routinely. Climbers will initially be delivered to the Marine Node as cargo. They will then be mated to the tether by the Operations team and loaded with customer payloads. The climber would then ride to the top under the control of the Operations center at HQ/POC. The current concept is a cylindrical shape enabling significant room for large and oddly shaped cargo.

2.2   The Tether's Mission

The mission of the tether is to provide a basic transportation mechanism for the movement of climbers and their payloads from the surface of the earth to LEO, GEO, and beyond. The mission includes returning payloads to disposal orbits or to the earth. The mission also includes returning payloads to disposal orbits or to the Earth.

2.3   Apex Anchor Mission

The Apex Anchor mission is multi-dimensional: but, its principle trait is to provide stability for the space elevator as a large end mass. This will ensure a firm tether for the climber, and provide a constant outward force. In addition, the Apex Anchor will have the mission of reeling the tether in and out as required for various tasks such as debris avoidance, damping tether end librations, and reacting to emergencies.

2.4   GEO Node Mission

The GEO Node mission is essentially the off-loading and on-loading of cargo to the tether climber. This zero-g location will facilitate delivery to the valuable Geosynchronous Ring of operational satellites. It will also be able to pick up payloads and return them to the earth's surface. A parallel mission would be to operate an assembly facility capable of fully enabling GEO satellites after their rise from the ocean. In addition, the GEO Node would allow interplanetary satellites to climb further and release toward solar system missions. Lastly, the GEO Node will be the primary component of space elevator communications architecture ensuring that operations at all of the nodes [GEO, Marine, Apex Anchor, and Climber] will be connected to the Headquarters and Primary Operations Center at all times.

2.5 Marine Node Mission

The mission is historic: support two-way transportation of goods while on the surface of the ocean. Basically, the Marine Node provides a location for the tether terminus that can enable safe and routine operations. This would include stabilizing the tether, moving the tether, loading and unloading of cargo, and local operations support. It is where the climber is prepared and then "sent on its way" safely. The Marine Node would tie together all of the aspects of the terrestrial component to include safety, security, inspection of cargo, loading of cargo to climber, loading climber to space elevator tether, off-loading climbers, and support to teams in the area.

3 Mission Philosophies and Strategies

This section will grow as the design of the tether and climber mature. The over-arching philosophy will be to ensure we have sufficient space elevators to avoid ever again having to use rockets for delivery of payloads to earth orbit. Permanently escaping the gravity well is a basic strategy of the space elevator development team. Figure 2 below depicts the Operations View (OV) 1 for the Space Elevator. Note that only one tether is shown.

Safety of humans, tether, climber, and existing orbiting satellites will be a constant concern. Strategies will evolve based upon the number of tethers (ascending and descending), and their capacities, (i.e., the number of simultaneous climbers). There will be strategies and constraints regarding direction of travel on tethers and the mix of LEO and GEO payloads. There will also be strategies for acquiring sufficient mass for the Apex Anchor which might include climbers. One concept that is being validated is collection of dead satellites from old GEO missions and raising them to the Apex Anchor on climbers designed to stay at the end of the tether.

The original system proposed by Dr. Edwards in THE SPACE ELEVATOR [SE] called for lasers to power climbers on their ascent to GEO. Powering all the way to GEO adds significant technical, political, and economic challenges to space elevator systems and operations. While lasers remain a potentially viable power source for elevator climbers, they are not available as a complete operational solution at this time. Given the economics of laser power in space elevator operations under the current perspective, this study chose to focus on an investigation without lasers for climber power. When suitable lasers and their operations become available, perhaps for laser launch, operations concepts for the space elevator will certainly be reevaluated.

Operation of climbers in the earth's atmosphere (first 30-40 km) will not be possible with solar arrays. The climber can be delivered above that altitude through many potential methods, such as: a Carbon Nanotube extension cord, lift by balloons, and High Stage One [discussed later].

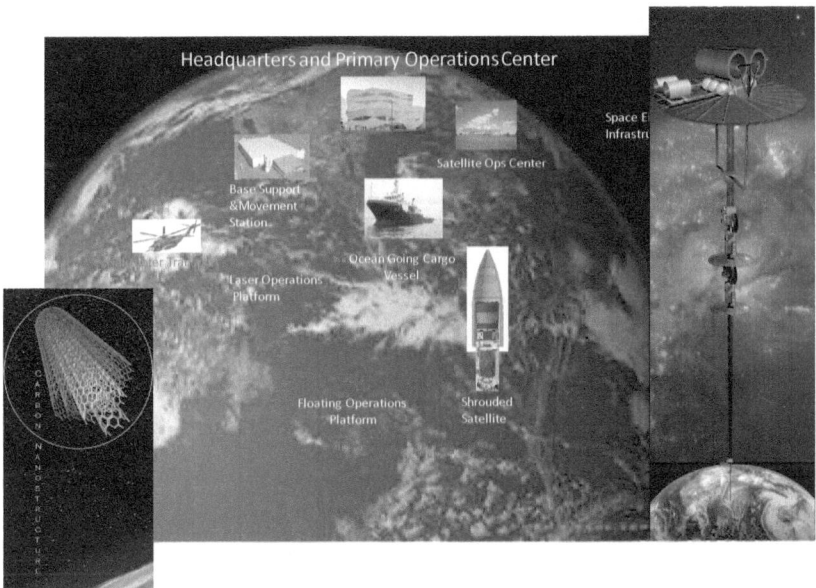

Figure 2   Space Elevator OV-1

4   Program Constraints

This section will also grow as the design of the tether and climber mature. Clearly, there will be a need to generate sufficient revenue to cover operation and maintenance (O&M) costs and to repay any loans for initial implementation. The full system of systems O&M costs will include replacement tethers and climbers.

Part II – System Characteristics

5   End-to-End Systems Components

5.1   Climber

The climber is the entity that ascends the tether while carrying a payload. Once outside the atmosphere, the climber is powered by solar cells. The tether climber will be fully instrumented and send health, status, and position telemetry to the ground based Telemetry Tracking & Command facility. The climber will provide power to its payload and can relay payload health and status information. It can climb at rates from meters per second to tens of meters per second. It may have multiple gears and

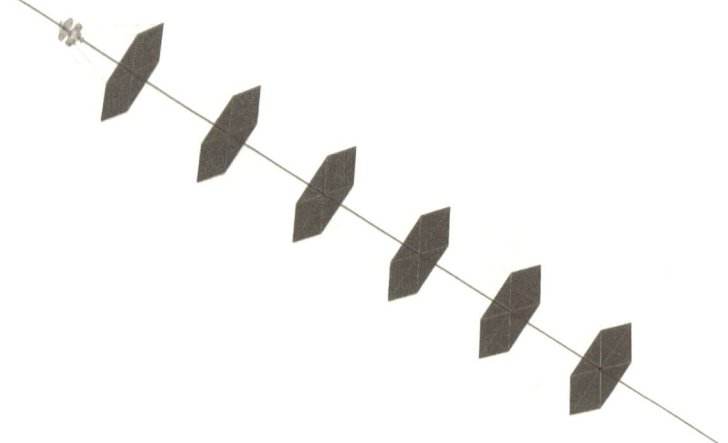

Figure 3 Climber with Solar Arrays [chasedesignstudios.com]

will have a robotic arm. Initial thinking is that it will weigh 6 metric tons and be able to lift 14 tons of cargo. Some climbers will carry tether repair apparatus and will execute repairs on the ascent, as needed. Figure 3 shows a climber with solar arrays. Figure 4 is a closer view of the climber.

Figure 4 below shows the layout of the Climber.

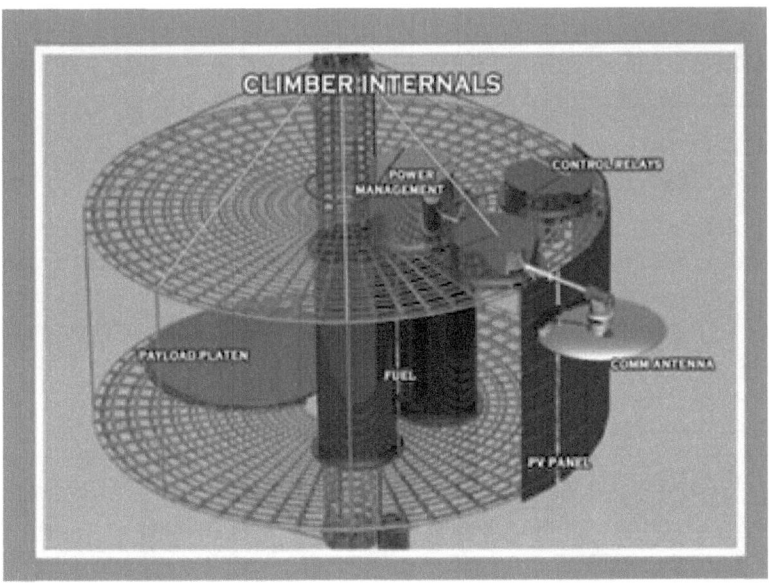

Figure 4 Climber Layout [Chasedesignstudios.com]

5.2   Tether

This carbon nanotube entity is about one meter wide at the bottom, a few mils thick, with a taper ratio of six, and approximately 100,000 km long. The space elevator will have one end fixed to the Marine Node while extending to an Apex Anchor. Tether/ribbon operations will be conducted mostly by operational personnel at the HQ/POC. Their principle responsibility will be to know the location and expected motion of each element of the space elevator. The requirement is to understand how to adapt the tether's natural motions for the operational needs such as climber motion, initiation of climb, avoidance of space debris, and motion around the GEO node. In addition, the team will monitor the health of the tether and schedule repair functions to be carried out including the "splicing" that might be necessary for construction of additional tethers. The strength of the tether determines how many climbers can be on it at any given time.

5.3   Apex Anchor

Operations of the Apex Anchor are controlled at the HQ/POC. This Anchor will have a lot of activity during construction with minimum activity once commercial operations commences. There may be some center of mass management that will result in the tether being reeled in or out of the Apex Anchor. In addition, the motion for avoidance of space debris could be controlled/ initiated from the Apex Anchor.

5.4 GEO Node

This element of the space elevator infrastructure will be the in-space location where much of the robotic and, in later years, human activity will be conducted. Off-loading and on-loading of cargo to the tether climber will be conducted at this zero-g location to facilitate delivery operational satellites to the Geosynchronous Ring of operational satellites. The assembly facility will be capable of fully enabling satellites after their rise from the ocean to geosynchronous altitude. The satellites will then be "sent off" to their operational location using either tugs or by their own power so they can begin their missions in the appropriate location. This geosynchronous station could be used to off-load satellites going to the planets, initiate them for flight, on-load them back to the space elevator and then watch them until they depart the climber into their high energy trajectories. A minor task is the de-mating of climbers from the tether and mating with a ferry for transportation to another tether on which they will descend. The GEO node will also receive spent satellites to be used as mass for the Apex Anchor. In addition, dead satellites could be mated with the climber for release into disposal orbits or return them to earth. It is also logical that the GEO Node becomes a refueling location as the price to raise thruster fuel is low and the infrastructure is safe. Operation of the GEO Node will leverage GEO satellite command and control functions that have been in practice since the early 60's

5.5 Marine Node

This is the virtual city on multiple floating platforms in the eastern Pacific ocean. The Floating Operations Platform (FOP) will be the size of an aircraft carrier or large oil tanker. Its primary purpose is to function as a terminus for supporting tether mating and de-mating of satellites and climbers. The Marine Node will have living quarters, kitchens and laundry, as well as recreational and medical facilities. It supports helicopter landings and loading/unloading from ocean going vehicles.

The FOP hosts a local Operations Center for management of tether, tether terminus, and platform operations. In addition, the center supports climber operations including operations/maintenance of the tether. The FOP can be a deep sea drillship which comes in two types: 1) It can be a ship which was designed and built to be a drilling vessel; or 2) It can be an older vessel which has been refitted with drilling equipment, or in our case, refitted to perform the functions of the FOP described above.

Drillships are self-propelled, carrying a complete ship's crew while underway, as well as a crew of drilling personnel-operations personnel in our case. Drillships are stabilized by either a standard anchoring system or by dynamic positioning of the vessel. Dynamic positioning is the use of a computer-operated inboard thruster system which keeps the vessel on location without the use of anchors. Examples of drillships are shown in Figure 5 below. The FOP will be painted in red and white checkerboard to enhance visibility. Its position will be reported to oceanic and space centers to diminish the probability of collisions. It will also have audible and visible warning lights and a keep-out zone for safety and security. In addition, the location and activity of the Marine Node will be maintained and broadcast to all aeronautical organizations to ensure flight safety.

Figure 5 Drill Ship Examples

5.5.1 High Stage One

This stage (if/when adopted) is a component of the Marine Node: but it is above it by 40 kms. It could be the terminus for the tether. The ability to attach the climber to the tether above the atmosphere leverages many factors, principally the need to operate in an environment not hostile to solar arrays. Large, fragile solar array structures cannot survive travel through the winds of the atmosphere: so, historically, they were stored and then deployed at altitude. The energy would immediately be available, at sunrise, for the tether climber to power the engine and rise from this Stage on its flight to GEO and beyond.

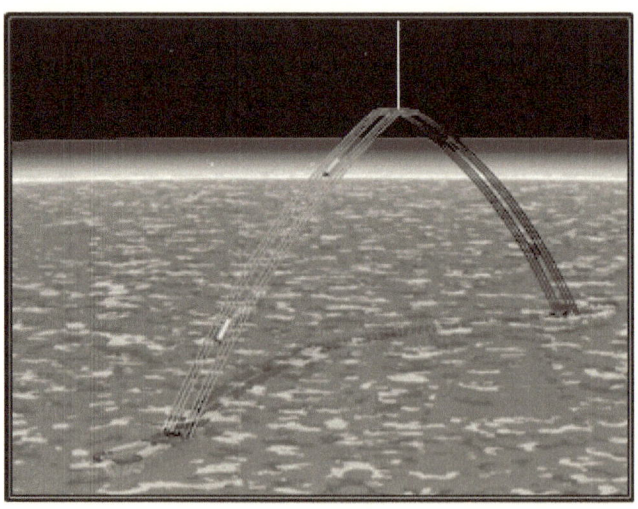

Figure 6   Lofstrom Loop

The concept is simple: place the working end of the space elevator on a firm platform at altitude. This facility would be capable of supporting 400 tons at 40 kms altitude with NO forces on the space elevator.

This transfer of hazards from the lower portion of the space elevator infrastructure to the terrestrial based Lofstrom Loop simplifies the problem. The Lofstrom Loop, shown in Figure 6 above, ensures stability of the platform at altitude and provides routine access from the ocean surface to 40 km altitude using electrical elevators on cable hoists available today.   Once the platform has been established at 40 kms altitude and the logistics "train" has geared up, the space elevator infrastructure becomes safer and simpler. Figure 7 shows the High Stage One platform at altitude. The current concept is a large operational building to protect the operators at altitude from radiation and to provide an atmosphere with pressure. Approximately five operators will be required at altitude; one crew chief, three loaders and one maintainer. As such, stresses on the space elevator become only space oriented forces, not atmospheric forces.

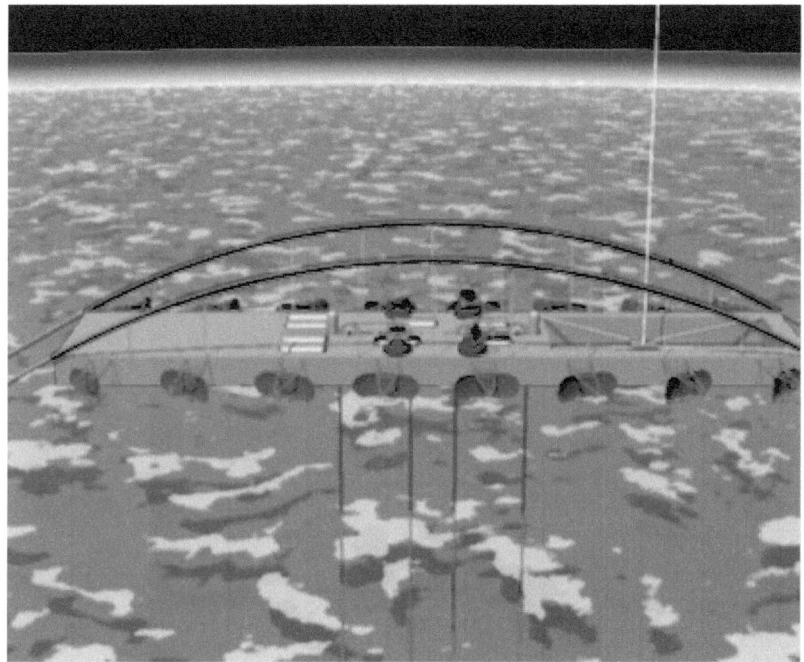

Figure 7 High Stage One Platform

5.6   Ocean Going Vehicle

There will be multiple vessels for the transport of satellites, climbers, equipment, supplies, and personnel to and from the Marine Nodes. When the second space elevator tether is deployed, this will include picking up climbers at the descending tether FOP and delivering them to the ascending FOP.

Figure 8 Ocean Going Vehicle

Occasionally, recovered satellites will be picked up for return to the Base Support Station. The ocean going vehicle must be capable of the round trip (approximately 5 days each way at 30 knots) to the Marine Nodes without refueling. Most likely, OGVs will be owned and operated by a vendor. Figure 8 above is a picture of the type vessel used today to supply offshore drilling rigs.   Helicopters will be used for the emergency transportation of equipment and personnel. The trip from the BSS to the FOP will likely require an intermediate re-fueling stop for aircraft. Helicopter services would be rented as needed.

5.7   Space Going Tug

A Space Tug will be needed for multiple purposes. Primarily, there is a need to re-orbit LEO satellites from the high apogee orbit resulting from release from the tether below GEO. The release could result in the desired perigee and inclination. More often, inclination will be adjusted, and then apogee and perigee adjusted to the desired orbit. The tug to accomplish this would be re-usable meaning. It would return itself to a climber on the tether between events. Tugs will also be needed to retrieve non-cooperative satellites for return to disposal orbit or earth. There could be a need to position GEO satellites after release from the tether. Most often, self-positioning would occur as that capability is inherent in station keeping requirements with most GEO satellites today. The space going tug would be refueled multiple times while in orbit to eliminate any need for frequent returns to the surface. Its parking place could easily be at the GEO Node with movement up and down the space elevator to meet its customer's potential needs.

5.8 Communications Architecture

The space elevator system will have a communication architecture that supports all facets of operations. This would include a communications hub at the GEO node with direct connections to all elements of the space elevator system. Marine Node facilities communicate to the tether climber early in the climb, but only until it escapes the atmosphere. The communications links are then set up with the GEO node and the tie into all other elements is assured. The GEO node connects to the FOP and to the HQ/POC with high through-put links. In addition, all other elements of the space elevator system can connect through these links to the GEO node and tie in with customers, satellite operators, elevator climbers, FOPs, ocean going ships, and personnel around the world. Figure 9 shows the communications architecture.

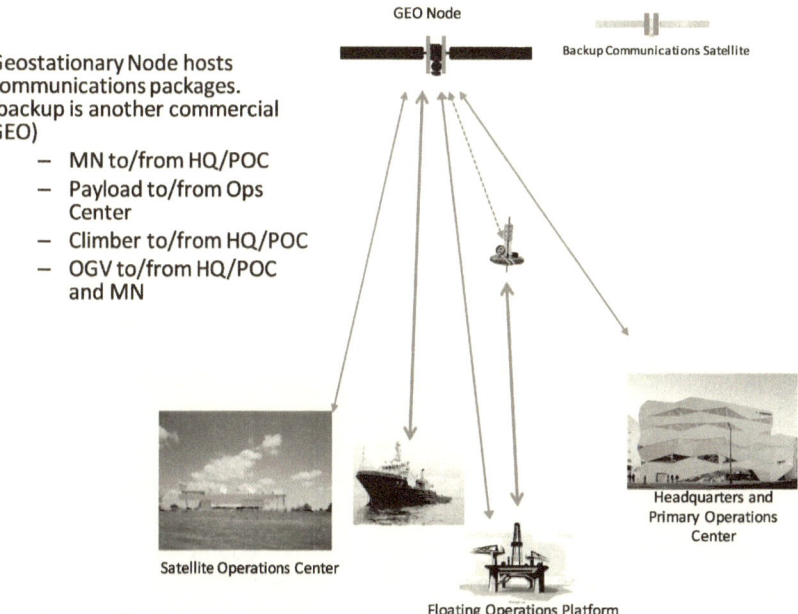

Figure 9 Communications Architecture

Part III – System of Systems Operations

6 Space Elevator System Operations

The operation of the total system of systems requires multiple locations where people are working to fulfill the missions of the space elevator. The table below shows the various operations centers and their locations.

Function	Location
Enterprise Operations Center	HQ & Primary Ops Center
Transportation Operations Center	HQ & POC
Climber Operations Center	HQ & POC
Tether Operations Center	HQ & POC
GEO Node Operations Center	HQ & POC
Marine Node Operations Center	Marine Node
Payload (Satellite) Operations Center	Owner's Ops Center

6.1 Headquarters and Primary Operations Center

The HQ/POC will host the key elements of conducting the business of transporting payloads to and from space. The business side will be the Enterprise Operations Center while the day to day execution of activities will be segmented to the various operations centers co-located within the HQ/POC. The HQ will represent the corporation while the POC will consolidate the operational functions of the system of systems. The following figure shows the layout of the facility with the table showing the significant functions to be handled by the HQ/POC. This co-location is by design as two factors will dominate: (1) the communications architecture will allow 24//7/365 connectivity to anywhere in the infrastructure and (2) co-locations should maximize efficiencies and minimize staffing demands. These operational activities will be conducted remotely, such as at the Marine Node, or in the future at the GEO Node. The functions to be accomplished at the HQ/POC, in addition to at the co-located operations centers, are:

- MP-Mission Planning
- APD-Activity Planning and Development
- MC-Mission Control
- DTD-Data Transport and Delivery
- NPA-Navigation Planning and Analysis
- SPA-Spacecraft Planning and Analysis
- PPA-Payload Planning and Analysis
- PDP-Payload Data Processing
- AMMD-Archiving and Maintaining the Mission Database
- SEIT-Systems Engineering, Integration and Test
- CCS-Computers and Communications Support
- DMS-Developing and Maintaining Software
- MMO-Managing Mission Operations
- FM-Financial Management

Headquarters and Primary Operations							
	Headquarters			Primary Operations Center			Likely co-located
	Corporate HQ	Transport Operations Center	Enterprise Operations Center	Climber Operations Center	GEO Node Operations Center	Tether Operations Center	Base Support Station
MP				x	x	x	
APD		x	x				x
MC				x	x	x	
DTD			x				x
NPD				x	x	x	
CPA				x			
PPA				x	x		x
CDP				x		x	
AMD			x				
SEIT	x		x				
CCS	x		x				
DMS	x		x				
MMO			x	x	x	x	
FM	x		x				

The HQ/POC can be located anywhere; but, for the initial concept it will be located in the greater San Diego, California area. It will have communications to all the other elements and will have an operations center staffed 24/7/365. Other sites closer to the Marine Node will be studied. Candidate sites must have both an international airport and a port on the Pacific Ocean. Stability of the government and overall security for personnel, equipment, and facilities will also be factors. The following sections discuss the various operations centers within the HQ/POC, address distribution of the operational activities, and look at staffing required.

6.1.1 Enterprise Operations Center

This infrastructure is the home for all the business operations as well as the administrative and logistics functions necessary for supporting the operation of the Space Elevator infrastructure. This will be the location for the lead on all financial transactions [Financial Management function] within the corporate infrastructure throughout the various facilities and centers spread around the world. The corporation headquarters is located at the HQ/POC to ensure day-to-day cognizance of the space elevator's business environment. The Enterprise Operation Center will focus on the revenue and expenses for the operations across the corporations. Operations will range from strategic planning for the enterprise to the research and development needs of future implementations. Day-to-day operations across the enterprise will be looked at; however, the functions will be to support sound business principles while conducting a transportation business.

6.1.2   Base Support Station

If not co-located with the Headquarters, this will be the forward support base for operations. Its focus will be on processing supplies, satellites (climber payloads), and climbers for transportation to/from the Marine Node. This will probably be located at a port for loading purposes. Staffing estimates are included with the HQ/POC estimates.

6.1.3   Transportation Operations Center

All transportation aspects of the enterprise will be controlled from the TOC. This is where payloads and climbers will be tracked. Location and status information will be monitored from the factory to the BSS to the MN, and then followed up the space elevator. Monitoring of returned payloads and climbers will also be done here. Arrangements for the ocean going vehicles will also be conducted here as well as planning for air transportation.

6.1.4   Climber Operations Center

This is where the majority of tether climber operations are conducted. Many of its activities include plans for delivery and maintenance of the various tether climbers that are on the tether(s) and in the process of being attached to the tether, or unattached to the tether. It is likely that the center will keep track of all tether climbers that are raising cargo to altitude, docked at the GEO node, and descending. Climber operations will consist mainly of monitoring the health and status of the climber: rate of climb, temperature of the motor(s) and wheels, and other health and status data. The monitoring of climbers will commence upon attachment to the tether. MN personnel will have access to all telemetry being sent to the Climber Operations Center after attaching and before ascent begins. Initiation of climb will be directed by the Climber Operations Center. These operations will leverage satellite operations that have been conducted since the early 60's.

Some climbers will perform repair operations which will likely be a combination of autonomous and operator involved activities. The climber, with payload, will ascend at a rate of meters to tens of meters per second using energy from solar panels. This would enable a round trip of about two weeks, or less. During periods when solar power is not available, the climber will remain stationary. Batteries on the climber and/or the satellite may be used to enable communication with the FOP and to perform housekeeping tasks. Movement of the tether to avoid space debris might also require the climber to park. An additional function to be performed in support of climber operations is the release of instrumented balloons from the FOP to collect high altitude weather information to support the operations of climbers under 40 km.

The shroud on the satellite will be removed at the direction of the satellite owner at the desired altitude. The climber, likely using its own robotic arms, will assist in removing the protective covering of the satellite and positioning the satellite away from the tether so any required thrusting will not harm the climber or the tether. Such an arm would also be useful for receiving and securing satellites for return to the Earth. One concept for an arm (from Tethers Unlimited) is shown in Figure 9 below.

After deployment of LEO satellites, the climber will continue to climb to the GEO node. There, it will pick up payloads for return to earth and begin its descent.

When multiple space elevators are in operation, the climber will be demated from the tether and ferried to the other GEO node for the descending tether. It may be loaded with a satellite for return to the earth's surface, it may pick up one on its descent, or return empty.

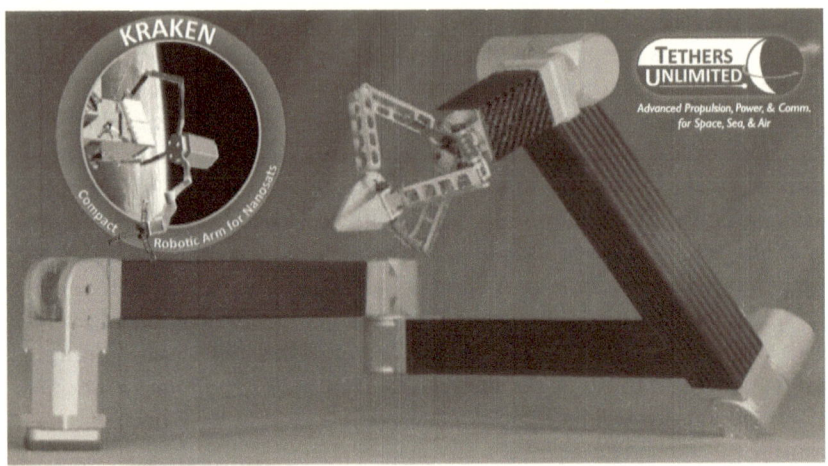

Figure 10   Robotic Arm Example

At this point, the climber may begin its return trip in the descending tether. In future operations, the climber may continue to higher altitudes with or without additional payloads and reach altitude where it will act as part of the apex anchor.

As the climber approaches the surface, it could deploy drag chutes to slow it down and brakes to bring it to a stop on the FOP. When arriving at the FOP, the climber will be de-mated for inspection and repair, as necessary. If another satellite payload is ready for lift, it will be mated to the climber and the climber will be mated to the tether. The satellite may be mated after the climber is mated to the tether. As the second space elevator becomes operational, returned climbers will be transported to the partner FOP.

6.1.5   Tether Operations Center

Knowledge of the three-dimensional location of all elements of a space elevators tethers [assume element is approximately five km long] is important to the operations of the total system of systems. Each tether climber location needs to be known continuously and monitored as to speed and expected location in the near future. This will enable the tether operations crew to understand their situation at all times. Concern for space debris impact becomes critical to successful operations through location maneuvering. This will consist mostly of operations personnel monitoring the probability of space debris impacting a tether. US Strategic Command (USSTR) will send out advisories predicting close approaches between large objects and the space elevator tether(s). Tether managers will decide whether to reel the tether in or out (and how much) to avoid possible collisions. Reeling out just a few meters of tether from the GEO host can impart tens of kilometers of lateral distance. Looking at an altitude of 660 km:

- 10 meters spooled out from GEO results in a little over 26 km lateral movement in the LEO region
- 100 meters spooled out results in about 83 km
- 1 km spooled out results in about 265 km

Managers will also decide whether or not to re-position after a predicted collision. Tether operations will include gathering of positional data for the tether(s) and reporting to USSTR. Note that debris avoidance can include re-positioning of the Marine Node terminus itself.

6.1.6   GEO Node Operations Center

The initial GEO node will have only robotic operations. As such, the operations center will be remoted to the HQ /POC. In the future when humans operate at the GEO Node, the operations center will be handling all tasks assigned to the large facility in GEO. Those functions include off-loading and on-loading payloads to climbers. In the future, refueling will be a principle mission at the GEO node as it is relatively inexpensive to deliver fuel rather than launch with it.

6.2   Marine Node [with Floating Operations Platform]

This is the virtual city of floating platforms in the eastern Pacific. The Floating Operations Platform will be the size of an aircraft carrier or large oil tanker. There will be one for each tether. Its primary purpose is to function as terminus for the tether while supporting mating and de-mating of satellites and climbers. The Marine Node will have living quarters, kitchen, laundry, recreational, and medical facilities. It supports helicopter landings and loading/unloading from ocean going vehicles. It will generate power with diesel generators. This may include power for the climber (through an extension cord) until it reaches 40 km where it can deploy its solar arrays. Alternative methods for the first 40 km are being studied. Operations would be handled by FOP personnel. The FOP hosts an Operations Center for local management of tether and platform operations through interaction with the HQ/POC operations center. This includes climber operations. A notional organization chart and estimated O&M Costs are shown later in this document.

6.3 Satellite [Payload] Operations Center

This is the satellite owner's Operations Center located at the customer's facilities. The satellite, while on the tether climber as a payload, will be in contact with the GEO node which will provide telemetry to the Satellite Operations Center. Commanding of the satellite will also be possible through this link.

6.4 Operations Staffing and Costing

Below is an Organizational Chart for the HQ/Primary Ops Center. The position descriptors have a direct correlation to experience, expertise, and education which would translate into most corporate personnel hierarchies.

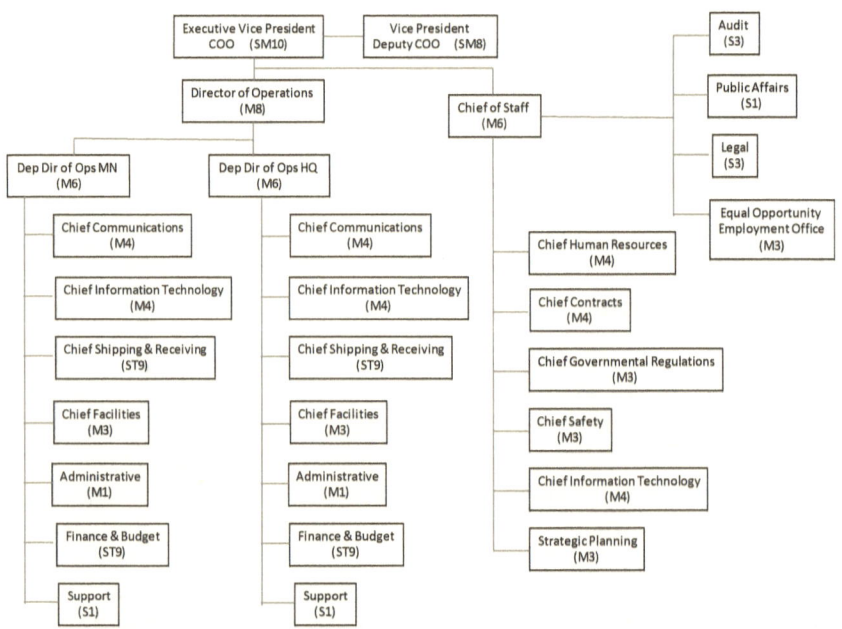

HQ/Primary Operations Center Organization Chart

The following charts show possible HQ/POC O&M costs and potential staffing for a single tether operation. The staffing will reflect a matrix organization with many operations functions interchangeable with each other. For example, all Climber Operations personnel would be qualified as Tether operations personnel. The next series of charts show the Marine Node information [Org chart, staffing, and O&M costs].

		HQ & Base Support Monthly Costs			
Fuel	2000				$2,000
Food	0				
Supplies	3000				$3,000
Comms	3000				$3,000
Building Lease	17000	Assume 100 sq ft per person plus 25% for larger offices for management and conference rooms			
Warehouse Lease	5000	Assume large enough to store 2 weeks of supplies plus 2 climber and 2 satellelites. Assume 5000 sq ft			
Single Lease	23000	per sq ft	100	$2,300,000	
Electric	6000				$6,000
Water	300				$3,000
Monthly					$2,317,000
Annually					$27,804,000
		To build our own building @___ persquare foot	250	$5,750,000	

HQ/POC O&M Costs

Executive Vice President COO					SM10	1	$250,000
	Executive Asst				ST8	1	$54,000
VP Deputy COO					SM8	1	$170,000
Chief of Staff					SM6	1	$110,000
	Dep COS				M5	1	$90,000
	Admin Asst				ST8	1	$54,000
	Audit				M3	1	$60,000
		Techs			ST7	2	$108,000
	Public Affairs				M1	1	$45,000
	Legal				M3	2	$120,000
	EEO				ST8	1	$54,000
	Human Resources				M4	1	$75,000
		Techs			ST7	2	$108,000
	Contracts				M4	1	$75,000
		Contract Specialists			ST7	4	$216,000
		Contract Administrators			ST7	4	$216,000
		Govermental Regulations			M3	1	$60,000
			Specialist		ST7	1	$54,000
		Safety			M3	1	$60,000
			Specialist		ST7	1	$54,000
		Information Technology			M4	1	$75,000
			Engineers		M3	2	$120,000
			Techs		ST7	4	$216,000
		Strtegic Planning			M3	1	$60,000
			Engineers		M1	2	$90,000

HQ/POC Staff

					Grade	Qty	Salary
Director Operations					O8	1	$170,000
	Dep Dir HQ				O6	1	$110,000
		Communications			O4	1	$75,000
			Specialists		E7	3	$162,000
			GEO SAT OPS		E7	3	$162,000
		Information Technology			O3	1	$60,000
			Techs		E7	3	$162,000
		Meteorology			O3	1	$60,000
		Shipping and Receiving			E9	1	$60,000
			Sr Techs		E7	2	$108,000
			Jr Techs		E4	4	$156,000
		Facilities			O3	1	$60,000
			Sr Techs		E7	5	$270,000
			Jr Techs		E3	5	$170,000
		Admin & Support			O1	1	$45,000
			Sr Techs		E7	3	$162,000
			Jr Techs		E3	5	$170,000
		Finance and Budget			E9	1	$60,000
			Specialists		E7	3	$180,000
			Specialists		E3	3	$102,000
		Climber Ops			O3	1	$60,000
			Engrs		O1	2	$90,000
			Techs		E7	2	$108,000
	Dep Dir FOP				O6	1	$110,000
		Communications			O4	1	$75,000
			Specialists		E7	3	$54,000
		Information Technology			O3	1	$60,000
			Techs		E7	3	$162,000
		Shipping and Receiving			E9	1	$60,000
			Sr Techs		E7	2	$108,000
			Jr Techs		E4	4	$156,000
		Facilities			O3	1	$60,000
			Sr Techs		E7	5	$270,000
			Jr Techs		E3	5	$170,000
		Admin			O1	1	$45,000
			Sr Techs		E7	2	$108,000
			Jr Techs		E3	2	$68,000
		Finance and Budget			E9	1	$60,000
			Specialists		E7	3	$162,000
			Specialists		E3	3	$102,000
		Support			O1	1	$45,000
			Sr Techs		E7	1	$54,000
			Jr Techs		E3	3	$102,000
			Med Tech		E7	1	$54,000
			Guard Sup		E8	1	$54,000
			Guards		E4	4	$39,000
						142	$7,534,000

HQ/POC Staff cont'd

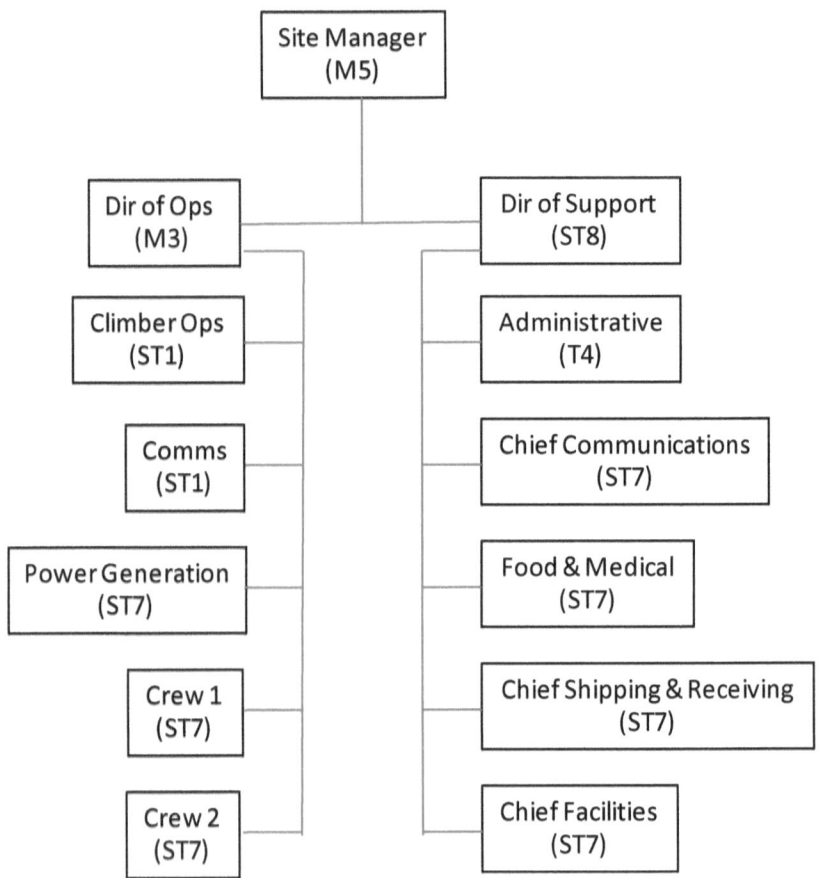

Marine Node Organization Chart

Below are the staffing estimates for the Marine Node using the same approach as the HQ estimates.

Site Manager			M5	1	$90,000	
	Admin/Support		T7	1	$54,000	
	Operations		M3	1	$60,000	
		Deck Officer	ST8	1	$204,000	
		Deck Hands	T5	2	$88,000	
		Training	ST7	0	$54,000	
		Trainee	T7	1	$44,000	
		Comms	ST7	1	$54,000	
		Comm Techs	T5	0	$0	
		Power Gen	T5	1	$44,000	
		Platform Maint	T5	6	$264,000	
	Support		ST8	0	$54,000	
		Food Prep	T5	3	$132,000	Shared among crew members
		Medical	ST6	1	$49,000	
		Laundry	T3	0	$0	Each does their own.
1	Crew				19	$1,191,000
2	Crews				38	$0
3	Crews				57	$3,573,000
			Total		$16,674,000	

Marine Node Staff

Summarizing the previous data leads to:

Yearly Operations Costs

HQ&POC Facilities O&M	28,000,000	build own facility for $ 6 million
HQ&POC staffing	7,500,000	
Marine Node (2) Facilities O&	30,400,000	
Marine Node (2) Staffing	33,400,000	
Total Yearly	$99,300,000	

6.5   Satellite Processing and Operations

6.5.1 Customer Payload Characteristics

Initial satellite designs are expected to perpetuate those for launch with rockets. Satellites designed for the space elevator environment with new materials will evolve. Traditionally, rockets have launched spacecraft into orbit. These satellites have been described in terms of a "bus" and a "payload." This construct works very well for the Space Elevator system in that we can refer to the climber as the bus carrying a payload. The climber was covered earlier. The payload is covered next. An important note on the climber is that it will have a large lift capability which will permit multiple payloads including classic "piggy back" payloads and others such as Get Away Specials from the Space Shuttle era. Satellite designs are expected to evolve to something noticeably different from today's satellites. The absence of "shake/rattle/roll" on launch, the abandonment of the 5 meter faring size restriction, and the hugely increased payload mass budget combine to mean satellite builders can now think of doing things they never seriously considered before. Below are thoughts on Space Elevator delivered payload characteristics in the new world of safe, reliable, easy ride, and routine environments.

6.5.2 Space Elevator Design Drivers

- **Shielding** – Designers can shield around critical components similar to the International Space Station protection against 1 cm size debris objects. Protecting against 10 cm sized objects would protect against over 90% of catalogued and un-catalogued objects. New lightweight radiation shields are also becoming available to ensure delivery of a payload that actually works in the space environment. Additional mass and volume capabilities allow spacecraft designers to achieve this level of shielding.

- **Additional Redundancy** – Designers can add redundancy to equipment to increase mean mission duration as the cost per kilogram to orbit is much smaller. This includes attitude control and propulsion to perform end of life (EOL) orbit lowering maneuvers or permit friendly capture by debris removal equipment. Mass to orbit cost reduction enables redundancy.

- **Additional Fuel** – Additional fuel would be available for orbit maintenance, avoidance maneuvers, and EOL maneuvers. This would also hedge against momentum wheel failures by permitting use of attitude control thrusters for 3 axis stability. Refueling, at greatly reduced cost of fuel would be standard.

- **Sensors** – Sensors could be added to help characterize many in-situ environmental characteristics, such as debris impacts and debris flux. This could extend to sensors which could provide space situational awareness observations.

- **Larger structure** – Larger structures would permit easy access panels for human or robotic on-orbit repairs, replenishment, and technology upgrades. In addition, larger structures could become routine vs. small confined spaces that astronautics have been forced into for the past 50 years. Large volume and high mass throughput will enable design expansion.

- **Capture Friendly features** – In addition , with the new expansion of volume constraints and mass limits, satellites could be built for more sophisticated debris mitigation techniques such as retrieval. This could include grappling points or actual mating hardware to permit easy capture for removal from orbit. This might also include "homing" devices to simplify the capture process.

6.5.3 Satellite Processing

Upon completion of System Integrations at the satellite builder's factory, the satellite will be placed in a container to provide protection from weather that would be expected to be encountered during transport from factory to San Diego. Upon arrival at the international airport, the satellite and any support equipment will be off loaded and transported to the Base Support Station. It will be stored until ready for transport to the Marine Node. The climber interface (power and T&C) may be used while in storage or while being transported to maintain its health and connectivity to its owner. It will be trucked from the BSS to the terminal for the ocean going cargo vessel (OGCV). It will then be loaded onto the OGCV, secured, and transported to the FOP in the eastern Pacific. Transit time will depend upon the OGCV's speed capabilities and the weather. The trip will take about 5 days if OGCV can manage 30 miles per hour. The payload container will be capable of being lifted by crane or forklift. It will also be wheeled for movement in the BSS and on the deck of the FOP. The satellite will remain in the container until it is ready to be deployed in space. Upon arrival at the FOP, the satellite will be lifted to the deck and wheeled to the location where mating with the climber will occur. Weather or other situations may make it necessary to wheel it into temporary storage. At the appropriate time, the satellite (in its container) will be mated to the climber. The climber may have been mated to the tether already. If not, the climber and mated satellite will be mated to the tether. Testing will occur to confirm climber and satellite are in the desired state including communications links from the FOP to the base station, the HQ/POC (if not co-located with the BSS), and, possibly, to the satellite owner's control center. Corrective actions will be taken as necessary. The climber will begin its ascent with a power cord attached. The cord will be designed such that it will not pose a hazard to the FOP or its personnel when disconnecting. The climber and payload will ascend at meters to tens of meters per second. During periods when power is not available, the climber will remain stationary ("parking brake set"). Batteries on the climber and/or the satellite may be used to enable communication with the FOP. At the desired altitude, the shroud on the satellite will be removed at the direction of the satellite owner. This may be accomplished by the satellite itself, autonomously, or under control from the SOC. Deployment from the container will be analogous to

38

deployment from the top of a rocket. One evolutionary path may lead to the situation where the climber is equipped with a robotic arm to perform shroud removal and satellite deployment. Such an arm would also be useful for receiving and securing satellites for a return to Earth.

At this point, the climber may continue to GEO or begin its return trip. From GEO, the climber may continue to higher altitudes with or without additional payloads It could reach an altitude where it might even be used to increase the mass of the counterweight.

When arriving at the FOP, the climber will be de-mated. It will be inspected and repaired, as necessary, then ferried to the ascending tether. If another satellite payload is ready for lift, it will be mated to the climber and the climber will be mated to the tether. (Note, the satellite may be mated after the climber).   Satellites will have the ability to interface with the climber's data up/downlink to provide Telemetry and Command capability for lift to orbital altitude. It will also be able to receive power from the climber.

There is no analog in the Space Elevator System for an End User. Clearly, the builders of the tether and the climber will be interested in getting feedback from tether operations (telemetry from the climber) on the performance of each component so the designs can be updated.

7   End User Data Products

There is no analog in the Space Elevator System for an End User. Clearly, the builders of the tether and the climber will be interested in getting feedback from tether operations (telemetry from the climber) on the performance of each component so designs can be updated.

Part IV – "A Day in the Life"

Joe is a supervisory Deck Officer on the FOP. He supervises operations on the deck including receiving climbers and supplies from the ocean going vehicle, mating climbers to tethers, mating payloads to climbers, managing the power cord to climbers for the first 40 km, and reel in/out of the tether for space debris avoidance. This day, Joe is on the FOP with the "up" climbers.

Joe works on a 28 day schedule with 5 days of travel each way on the OGV and 18 days of work on the FOP. On day 1, Joe boards the OGV for the ride to the FOP. The 5 days will be spent with some time each day for refresher training as well as training for new features and functions on the FOP. The remainder of his waking hours is free time. The OGV will have facilities for recreational activities including internet access. Access to a satellite phone is also available.

Upon arrival at the FOP, Joe and his crew receive a briefing from the off-going deck officer while the OGV is off-loaded by both the on-coming and off-going crews. In parallel, fuel is pumped to tanks near the generators. When everything is off-loaded, any cargo (returned satellites and climbers, failed FOP hardware/equipment) is on-loaded. Lastly, personnel are loaded and the OGV departs. The off-loaded items are delivered to the place of usage: food to the kitchen, supplies to the warehouse, satellites and climbers to the storage area and so on. When complete personnel then proceed to their quarters and stow their gear. Joe then proceeds to a conference room and meets with his crew. He relates any issues or concerns from the crew just relieved and briefs the schedule for the next 17 days. He then directs which individuals will perform in each of the positions. This includes assignments as training givers and training receivers. The rest of the day is free time. Joe may choose to do some laundry or he may prepare a meal for himself. He may watch a movie in his room or play cards in one of the group activity rooms or he may read a book. There will be two hour shifts for security detail shared between all the crew members including Joe. This means monitoring and responding security alarms.

Joe will experience different kind of work days while on the FOP. One type of work day is repair climber mating to the tether. One of Joe's crew releases a weather balloon to gather meteorological data. The balloon transmits its data to the Operations center which forwards back to HQ. The repair climber is retrieved from storage and mated with the tether. Today, they are using a new procedure they studied while on the OGV. There is also a trainee on the FOP for the first time, which is being shadowed by a qualified operator. While supervising the operation Joe is in contact with the FOP operations center. Mating includes connecting the power cord to the climber. The repair climber does not carry a payload so there is none to mate. The climber is fully instrumented and sends its health and status telemetry to the FOP and to the GEO Node which forwards to the HQ. Today's weather is acceptable so the climber begins its ascent under the control of HQ and monitored by FOP personnel. Joe and his crew monitor the climber while it's in view, monitor the reeling out of the extension cord, and monitor the movement of the tether. Once the climber is out of sight, all monitoring is done via telemetry in the HQ operations center. When the climber reaches 40 km, it "parks" and disconnects the power cord which is reeled back in by the FOP crew. The climber then deploys its solar arrays. When there is adequate sunlight, the climber will continue climbing.

Joe and his crew now begin routine maintenance activities. This includes preventive maintenance on the myriad FOP subsystems. It includes wire brushing and painting of most metallic surfaces and it includes washing the decks to remove salt water. The normal work day is 12 hours, approximately sunrise to sunset. The routine maintenance activities continue throughout each day with time taken for breaks, meals, and training sessions. After the evening meal is free time until sunrise.

Another type of day is a climber/payload scheduled to be launched. The activities are the same except that the payload must be retrieved from storage and mated to the climber.

Another type of work day is the return of a climber. This day includes monitoring the descent of the climber when it becomes visible. This activity would focus on the deployment of drag parachutes and brakes to slow the speed of the climber. When Joe is working on the ascending tether FOP, his work days are of the type described above minus the de-mating of climber/payloads from the tether. When Joe is working on the descending tether FOP, his work days are of the type described above minus the mating of climber/payloads to the tether.

The last type of work day is no climber/payloads scheduled to be launched or recovered. This type work day is spent in preventative maintenance and training activities.

Joe's remaining days on the FOP are spent experiencing one or more of each of these kind of work days. 19 days after arriving, Joe boards the OGV for the return trip to San Diego. The days are spent the same as the outbound trip. Joe will have time off when he gets back, nominally two weeks. He is also able to extend his time off by using vacation days. After his time off, he will either get on an OGV or report to HQ for duties as assigned.

8 Findings and Conclusion

Below are the Findings of this report:

- Operation of the Space Elevator will leverage over 50 years of experience in operating satellite systems. The Climber, Apex Anchor, and GEO node are essentially satellites. Space Elevator operations will be an easy extension of today's practices. Operations centers will look very much like today's satellite operations centers.
- Operation of the Space Elevator will leverage over a hundred years of experience in off-shore drilling operations. The Floating Operations Platform will likely be a modified drilling platform. Support to off-shore drilling platforms is a mature industry.
- The Operation and Maintenance Costs appear to be reasonable.

The Major Conclusion that resulted from this year-long study by the International Space Elevator Consortium is:

Operations for a Space Elevator
Have No Showstoppers
Have Reasonable Costs &
Meet the Challenge

9   References

CESMO – <u>Gael Squibb</u>, <u>Daryl Boden</u>, and <u>Wiley Larson</u>, ***Cost Effective Space Mission Operations***, McGraw Hill, 1996.

ASSE – <u>Wiley Larson, Doug Kirkpatrick, Jerry Sellers, L. Dale Thomas, and Dinish Verma</u>, ***Applied Space Systems Engineering***, McGraw Hill, 2009.

SES-SDM – <u>Peter Swan, Robert "Skip" Penny, and Cathy Swan, *Space Elevator Survivability – Space Debris Mitigation*</u>, Lulu.com, 2011.

SE – Brad Edwards and Eric Westling, ***Space Elevator – A revolutionary Earth-to-space transportation system***, BC Edwards publishing, 2002.

SE-SA – Peter Swan and Cathy Swan, ***Space Elevator Systems Architecture***, Lulu.com, 2007.

Appendix A   ISEC Mission

The International Space Elevator Consortium has the following Mission Statement:

> *...ISEC promotes the development, construction and operation of a space elevator as a revolutionary and efficient way to space for all humanity ...*

Development
Humanity's operations in space have taken many different forms. Individual governments, international partnerships and commercial ventures have all made their forays into the heavens. Which form is best for a Space Elevator? How can this great enterprise be used to its greatest potential to help all humanity? These are the questions that ISEC will try to answer.

Construction
History shows that having a great idea is not enough. Large endeavors such as the first trans-Atlantic telegraph cable or the building of the Panama Canal required immense support, often from just a few individuals, in order to push them to completion. Once the technology is available, ISEC wants to ensure that the technology is properly implemented.

Appendix B   Biographies

Robert E. "Skip" Penny, Jr. *graduated from the US Air Force Academy in 1970 with a Bachelor of Science degree. Over his 20 year Air Force career, he held a wide range of command and staff positions in NORAD/ADCOM, Air Force Space Command, US Space Command, and Air Force Technical Applications Center, retiring as a Lieutenant Colonel.*

Upon retirement in 1990, he joined Motorola on the Iridium satellite program. As a System Engineer, he initially provided operations input to the early Iridium system design including authorship of the Iridium System Operations Concept and the Control Segment Operations Concept. He was a key contributor to initial release and multiple updates to A level specifications and segment interface control documents. He generated multiple Iridium Technical Notes on operations related functions including a probability of collision assessment with recommendations for debris mitigation.

In 2000, he went to work for General Dynamics as Senior System Engineer. He was Network and Communications Integrated Product Team Lead for General Dynamics-Lockheed Martin GPS III System Engineering and Integration Team. He was responsible for system and segment level requirements and resulting design of GPS III's network of ground and space nodes including crosslinks.

*Skip has a Master of Science degree in Space Operations from the US Air Force Institute of Technology. His Masters thesis was a computer simulation that predicted the probability of collision for the US Space Shuttle using a methodology that has since been adopted by AIAA, and many space operators. He also has a Master's Degree in Procurement Management from Webster College. He is the co-author of **Space Elevator Survivability – Space Debris Mitigation***

Dr. Peter A. Swan *graduated from the US Military Academy in 1968 with a Bachelor of Science degree. Over his 20 year Air Force career, he held a variety of research and development positions in the space arena. He taught at the Air Force Academy and retired as a Lieutenant Colonel.*

Upon retirement in 1988, he joined Motorola on the Iridium satellite program. As a System Engineer, he initially provided space systems engineering leadership to the field office in Washington D.C. He then led the team responsible for the development of the Iridium spacecraft bus. His management of the Lockheed Martin team coupled with engineering insight into the hostile environment of space enabled Motorola to successfully launch 96 healthy spacecraft that are currently working 12 years past their 5 year design life.

In 1998, he developed his own company and taught space systems engineering for Teaching Science and Technology, Inc. His classes emphasize engineering know-how and management techniques to successfully develop space systems of national importance. He also taught Introduction to Space and Commercial Space Systems Engineering. He still teaches several times a year.

In 1996, Dr. Swan was invited to participate on the Air Force Scientific Advisory Board and contributed for five years attacking many of the Air Force's toughest issues. From there, he was asked to join the Army Science Board for another 10 years looking at their issues from a space systems engineering perspective.

Pete's final degree was a Ph.D. from the University of California at Los Angeles in Mechanical Engineering with a specialty in space systems. His dissertation was on tethered satellites. This led to a natural extension to an interest in space elevators where he has participated for over ten years with great interest. He has published many papers and a few books; two of which are on preparing for SCUBA trips. However, his space elevator publications include: **Space Elevators Systems Architecture** and **Space Elevator Survivability – Space Debris Mitigation.**

Dr. Cathy W. Swan graduated with a Ph.D. from the University of California at Los Angeles with a specialty in Space Policy. Her dissertation topic dealt with impact of long duration spaceflight and the stresses placed upon the travellers. A highlight of her Air Force career was a tour in the Management Department of the USAF Academy. She retired as a Colonel after 30 years of service.

In 1989 she established the Center for Arms Control and Technology Assessment in Washington D.C. to deal with the tremendously complex world of satellite operations and arms control and treaty verification. She has continued to be active in her career field with consulting and as President of SouthWest Analytic Network, Inc. She has authored many papers and books. Her latest books are: **The Impact of Space Activities upon Society, National Security Space Strategy Considerations,** Space **Elevators Systems Architecture** and **Space Elevator Survivability – Space Debris Mitigation.**

In 1994 she joined the Phoenix Art Museum as a docent where she does technical support for docent activities. She has since established herself as a Master Docent and the "go-to" person for questions on how to successfully operate technological displays of art.